Explain It To Me:

This publication is written to provide helpful information on how to use a Telescope.

David Laney

ISBN: 9798882533853

First Edition 2024.

Foreword

Are you looking for the perfect introductory guide on how to use a Telescope?

Look no further! In this publication, you'll find all the explanations on the use and history of the Telescope.

Dedication

I am dedicating this publication to my wife, Darleen, who has always supported me in this and every endeavor.

Table of Contents

Introduction

The cosmos exploration has captivated humanity's imagination for millennia, igniting a deep-seated curiosity about the vast expanse of the universe beyond our planet Earth. At the heart of this enduring fascination lies the telescope, a remarkable instrument that has revolutionized our understanding of the cosmos and unlocked the secrets of distant stars, galaxies, and nebulae. In this introductory narrative, we embark on a journey through the history, technology, and practical applications of telescopes, exploring why this topic is of such enduring interest to scientists, astronomers, and enthusiasts alike.

The development of telescopes has been driven by a relentless quest to explore the unknown, to peer deeper into the cosmos, and to unravel its mysteries. From the invention of the refracting telescope by Dutch spectacle-maker Hans Lippershey in the early 17th century to the launch of the Hubble Space Telescope in 1990, each milestone in telescope technology has brought us closer to the stars, revealing the beauty and complexity of the universe in ever greater detail. Today, telescopes come in various shapes and sizes, from portable refractors to massive radio telescopes spanning acres of land, each offering unique insights into the cosmos.

But why are telescopes so fascinating? What is it about these humble instruments that continue to captivate our imagination and inspire wonder in people of all ages and backgrounds? At its core, the appeal of telescopes lies in their ability to transcend the limitations of human vision to extend our gaze beyond the confines of Earth and into the depths of space. With a telescope, we can see farther, more precisely, and with greater detail than ever before, unlocking vistas of the universe that were once unimaginable.

Telescopes serve as portals to the past, allowing us to glimpse the light of distant stars and galaxies as they appeared millions or even billions of years ago. By studying this ancient light, astronomers can piece together the history of the cosmos, tracing the evolution of galaxies, the formation of stars, and the birth of planets. Through telescopes, we can explore the universe's wonders in all their majesty, from the fiery birth of stars in distant nebulae to the violent collisions of galaxies billions of light-years away.

But telescopes are not just tools for scientific discovery; they are also windows to the sublime, connecting us to something greater than ourselves and inspiring a sense of wonder and humility in the face of the vastness of the cosmos. Whether we're observing the delicate rings of Saturn, the swirling clouds of Jupiter, or the distant glow of a newborn star, telescopes remind us of our place in the universe and our shared journey through space and time.

In this publication, we will explore the history of telescope development, from its humble beginnings to the cutting-edge technology of modern observatories. We will delve into the different types of telescopes, from refractors and reflectors to compound designs, examining their strengths, weaknesses, and applications. We will also discuss the essential features of telescopes, including aperture, focal length, and mount type, and how these factors impact their performance and versatility.

Introduction

Observing techniques will be a crucial focus of our discussion as we explore the art and science of observing the night sky with a telescope. From naked-eye observation and star hopping to using finder scopes and star charts, we will uncover the secrets of navigating the celestial landscape and locating elusive celestial objects. We will also discuss practical tips for setting up and maintaining a telescope, from assembly and alignment to collimation and storage, ensuring that your instrument remains in peak condition for years.

Throughout our exploration, we will encounter the wonders of the night sky, from the familiar constellations of the northern hemisphere to the breathtaking beauty of distant galaxies and nebulae. We will learn how to identify significant stars and constellations, locate prominent deep-sky objects, and observe planets and other celestial bodies with clarity and precision. We will also discuss the problem of light pollution and the importance of dark sky sites for preserving the natural beauty of the night sky and reducing its impact on astronomical observation.

The study of telescopes is a journey of discovery and exploration that transcends the boundaries of space and time. From their humble beginnings to the cutting-edge technology of today's observatories, telescopes have revolutionized our understanding of the cosmos and inspired generations of astronomers, scientists, and enthusiasts to explore the universe's wonders. Whether you're a seasoned astronomer or a novice stargazer, the telescope offers endless opportunities for discovery, learning, and inspiration, unlocking the mysteries of the cosmos and revealing the beauty of the universe in all its glory.

Introduction

History of Telescopes

The history of telescopes is a journey through time marked by the ingenuity and curiosity of astronomers and scientists. It all began in the early 17th century when Galileo, an Italian polymath, turned his telescope skyward. Galileo's telescope, although rudimentary by today's standards, revolutionized our understanding of the cosmos. With it, he made groundbreaking discoveries, including the moons of Jupiter and the phases of Venus, challenging the geocentric model of the universe and laying the foundation for modern astronomy.

Following Galileo's pioneering work, astronomers and instrument makers worldwide began refining and improving the design of telescopes. In the 17th and 18th centuries, notable figures like Johannes Kepler and Isaac Newton contributed to the development of refracting and reflecting telescopes, respectively. Kepler's work on optics led to the design of the Keplerian telescope, which corrected many of the flaws in Galileo's original design. Meanwhile, Newton's invention of the reflecting telescope, with its curved mirror instead of a lens, overcame many of the limitations of refracting telescopes, such as chromatic aberration.

Introduction

Throughout the centuries, telescopes continued to change, driven by optics, materials science, and engineering advancements. The 19th century saw the rise of large refracting telescopes, such as the famous Great Refractor at the Yerkes Observatory, which boasted a massive 40-inch aperture. Meanwhile, the development of the Cassegrain and Newtonian designs in the 17th and 18th centuries laid the groundwork for modern compound telescopes, combining refractors and reflectors' best features.

The 20th century brought about revolutionary changes in telescope technology with the advent of computerized telescopes and space-based observatories. The launch of the Hubble Space Telescope in 1990 marked a new era in astronomy, providing unprecedented views of the cosmos free from the distorting effects of Earth's atmosphere. In recent years, advances in adaptive optics and interferometry have further enhanced the capabilities of ground-based telescopes, enabling astronomers to study distant galaxies and exoplanets with unprecedented clarity and precision.

Today, telescopes come in all shapes and sizes, from portable refractors to massive radio telescopes spanning acres of land. They are used not only by professional astronomers but also by amateur enthusiasts and citizen scientists around the world. With each discovery and technological breakthrough, telescopes continue to expand our horizons, revealing the beauty and complexity of the universe in which we live.

Types of Telescopes

Telescopes come in various designs, each with its strengths and weaknesses. The three primary types of telescopes are refractors, reflectors, and compound telescopes, each offering unique advantages for different observing purposes.

1. **Refractor telescopes**, or dioptric telescopes, utilize lenses to gather and focus light. They have a simple and rugged design, making them well-suited for beginners and casual observers. Refractors provide crisp, high-contrast images of celestial objects, making them ideal for planetary observation and lunar photography. However, they are limited by their size and cost, as larger refractors can become prohibitively expensive.

2. **Reflecting telescopes**, on the other hand, use mirrors to collect and focus light. They come in various configurations, including Newtonian, Cassegrain, and Ritchey-Chrétien designs. Reflectors are prized for their larger apertures and lower cost per inch of aperture compared to refractors, making them popular among amateur astronomers and astrophotographers. They excel at deep-sky observation, revealing faint galaxies, nebulae, and star clusters invisible to the naked eye.

3. **Compound telescopes** combine the best features of refractors and reflectors, typically featuring a combination of lenses and mirrors in their optical path. The most common type of compound telescope is the Schmidt-Cassegrain, which uses a corrector plate, primary mirror, and secondary mirror to fold the light path, resulting in a compact and versatile instrument. Compound telescopes offer excellent image quality, portability, and convenience, making them popular for astrophotography and visual observation.

Each type of telescope has advantages and disadvantages, and the choice ultimately depends on the observer's preferences, budget, and intended use. Whether you're a beginner looking to explore the wonders of the night sky or an experienced astronomer seeking to push the boundaries of our understanding, there's a telescope out there to suit your needs.

Explain It To Me: Telescope

Basic Telescope Features

When selecting a telescope, several key features must be considered to ensure optimal performance and suitability for observing needs. These features include aperture, focal length, mount type, eyepieces, and accessories.

1. **Aperture** refers to the diameter of the primary optical component of the telescope, whether it's a lens or a mirror. It determines the telescope's light-gathering ability and resolution, with larger apertures collecting more light and revealing finer details in celestial objects. A larger aperture allows you to observe fainter objects and achieve higher magnifications, making it crucial for deep-sky observation and planetary imaging.

2. **Focal length** is the distance from the primary optical component to the focal point, where light rays converge to form an image. It determines the telescope's magnification and field of view, with longer focal lengths providing higher magnifications and narrower fields. On the other hand, shorter focal lengths offer more expansive fields of view and are better suited for observing large celestial objects like galaxies and nebulae.

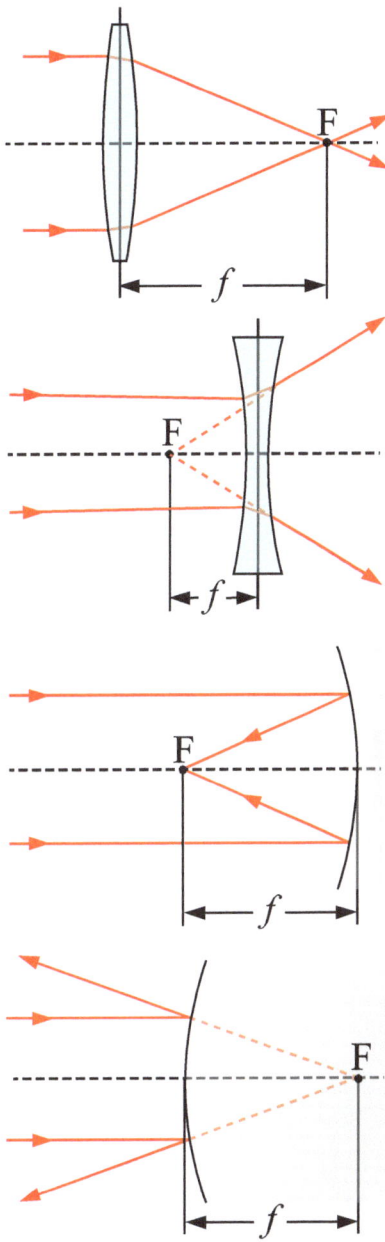

3. **Mount type** plays a significant role in the stability and tracking accuracy of the telescope. There are two primary types of mounts: alt-azimuth and equatorial. Alt-azimuth mounts move the telescope in vertical (altitude) and horizontal (azimuth) directions, making them intuitive and easy to use but less suitable for long-exposure astrophotography. Equatorial mounts, on the other hand, align with the Earth's axis of rotation, allowing for precise tracking of celestial objects as they appear to move across the sky. Equatorial mounts are preferred for astrophotography and advanced observing techniques like star hopping and manual monitoring.

4. **Eyepieces** and **magnification** are critical considerations for visual observation. Eyepieces come in various focal lengths and designs, each providing a different magnification and field of view. A range of eyepieces allows observers to tailor their viewing experience to suit celestial objects and observing conditions. Additionally, accessories such as Barlow lenses, filters, and focal reducers can enhance the telescope's versatility and performance, providing options for increasing magnification, improving contrast, and reducing image distortion.

Understanding these key features empowers observers to make informed decisions when selecting a telescope and accessories that best suit their observing goals and preferences. Whether you're a beginner exploring the night sky for the first time or an experienced astronomer seeking to push the boundaries of discovery, the right telescope can unlock a universe of wonder and exploration.

Telescope: Celestron AstroMaster

The Celestron AstroMaster Telescope is an ideal choice for amateur astronomers and enthusiasts eager to explore the wonders of the night sky. Among its offerings, the Celestron AstroMaster with eyepiece kit stands out as a popular option, combining a robust optical design with essential accessories to provide users with a comprehensive observing experience.

At the heart of the Celestron AstroMaster Telescope is its optical system, meticulously crafted to deliver crisp, clear views of celestial objects ranging from the moon and planets to distant galaxies and nebulae. The telescope features a refracting optical design, utilizing precision-crafted lenses to gather and focus light with exceptional clarity and sharpness. With an aperture ranging from 70mm to 130mm, depending on the specific model, the AstroMaster Telescope offers ample light-gathering power to reveal intricate details and subtle features in the night sky.

One of the standout features of the Celestron AstroMaster Telescope is its versatile mount, which provides stable support and smooth tracking for easy navigation of the night sky. Depending on the model, the AstroMaster may come equipped with an alt-azimuth or equatorial mount, each offering unique advantages for different observing techniques and applications. The alt-azimuth mount is intuitive and easy to use, allowing users to pan and tilt the telescope in any direction with effortless precision. Meanwhile, the equatorial mount aligns with Earth's axis of rotation, facilitating accurate tracking of celestial objects as they appear to move across the sky. Both mounts feature sturdy construction and ergonomic controls, ensuring a comfortable and enjoyable observing experience for users of all skill levels.

To complement its optical performance and mount design, the Celestron AstroMaster Telescope has a comprehensive eyepiece kit, providing users with various magnification options to suit different observing conditions and targets. The eyepiece kit typically includes a selection of high-quality eyepieces with varying focal lengths, allowing users to adjust the telescope's magnification to achieve optimal views of celestial objects. The kit may include accessories such as a Barlow lens, which doubles or triples the telescope's magnification, and filters to enhance the contrast and visibility of specific features in the night sky.

Setting up and using the Celestron AstroMaster Telescope is straightforward, thanks to its user-friendly design and intuitive controls. Whether you're a beginner exploring the night sky for the first time or an experienced astronomer seeking to expand your horizons, the AstroMaster Telescope offers a hassle-free observing experience that encourages curiosity and exploration. Simply assemble the telescope according to the included instructions, align the mount with the celestial pole or a known landmark, and you're ready to embark on a journey through the cosmos.

Telescope: Orion SkyQuest XT8 Dobsonian

The Orion SkyQuest XT8 Dobsonian telescope is a perennial favorite among amateur astronomers, renowned for its affordability, simplicity, and exceptional optical performance. Featuring a classic Dobsonian design, this telescope offers an impressive 8-inch aperture, providing ample light-gathering power to reveal stunning views of the moon, planets, and deep-sky objects.

At the heart of the SkyQuest XT8 is its high-quality parabolic primary mirror, meticulously crafted to deliver crisp, clear images with minimal distortion and aberration. This mirror is complemented by a sturdy wooden alt-azimuth mount, which provides smooth, effortless movement and allows users to quickly point the telescope at celestial objects of interest.

One of the standout features of the SkyQuest XT8 is its simplicity and ease of use. Unlike complex computerized telescopes, the XT8 requires no alignment or setup procedures—simply point the telescope at the night sky and start observing. This makes it an ideal choice for beginners and seasoned observers alike, allowing users to focus on the beauty of the cosmos without the distractions of complicated technology.

The SkyQuest XT8 also offers versatility in its observing capabilities. With its generous aperture and stable mount, it excels at planetary and deep-sky observation, revealing intricate details on the surfaces of planets like Jupiter and Saturn while also providing breathtaking views of distant galaxies, nebulae, and star clusters.

Telescope: Sky-Watcher 10-inch Collapsible

The Sky-Watcher 10-inch Collapsible telescope builds upon the success of its smaller counterparts, offering amateur astronomers a larger aperture and greater light-gathering power for enhanced night sky views. Featuring a collapsible truss-tube design, this telescope combines impressive optical performance with portability and convenience, making it an ideal choice for observers on the go.

With its 10-inch aperture and parabolic primary mirror, the Sky-Watcher Collapsible Dobsonian provides breathtaking views of celestial objects with exceptional clarity and detail. From the intricate cloud bands of Jupiter to the delicate filaments of distant nebulae, this telescope reveals the universe's wonders in stunning detail, inspiring awe and wonder in observers of all skill levels.

Despite its larger aperture, the Sky-Watcher Collapsible Dobsonian remains remarkably lightweight and easy to transport, thanks to its innovative truss-tube design. This collapsible construction allows the telescope to be disassembled into smaller, more manageable components, making it easier to transport and set up for observing sessions under dark skies.

Like its smaller counterparts, the Sky-Watcher Collapsible Dobsonian features a sturdy alt-azimuth mount that provides smooth, precise movement and allows users to easily track celestial objects as they move across the night sky. With its intuitive controls and user-friendly design, this telescope offers a hassle-free observing experience that encourages exploration and discovery.

In conclusion, the Sky-Watcher 10-inch Collapsible Dobsonian telescope is a versatile and powerful instrument that delivers exceptional performance at an affordable price. Whether you're a beginner exploring the night sky for the first time or an experienced observer seeking to push the boundaries of discovery, the Collapsible Dobsonian offers a gateway to the wonders of the cosmos, inspiring awe and wonder in all who gaze through its eyepiece.

Telescope: Explore Scientific ED102 Refractor

The Explore Scientific ED102 Refractor telescope represents the pinnacle of optical performance and craftsmanship, offering amateur astronomers stunning night sky views with exceptional color correction and sharpness. Featuring a high-quality apochromatic lens system, this refractor telescope delivers breathtaking images of celestial objects with minimal chromatic aberration and distortion.

At the heart of the Explore Scientific ED102 is its precision-crafted ED (extra-low dispersion) glass lens, designed to reduce chromatic aberration and produce high-contrast, true-color images of celestial objects. This lens is complemented by a sturdy yet lightweight aluminum tube assembly, which provides excellent stability and support for the optical system.

One of the standout features of the Explore Scientific ED102 is its versatility and adaptability for both visual observing and astrophotography. With its 102mm aperture and 714mm focal length, this telescope offers ample light-gathering power and magnification for observing a wide range of celestial objects, from the moon and planets to distant galaxies and nebulae. Whether exploring the Milky Way's rich tapestry or capturing stunning images of distant galaxies, the ED102 delivers exceptional performance that inspires awe and wonder in observers of all skill levels.

The Explore Scientific ED102 also features a precision-crafted Crayford-style focuser, which provides smooth, precise focusing and allows users to quickly achieve sharp, high-contrast views of celestial objects. With its dual-speed design, this focuser offers fine adjustments for critical focusing and coarse adjustments for rapidly targeting objects in the night sky.

Telescope: Meade LX90 ACF

The Meade LX90 ACF telescope represents the pinnacle of optical performance and technological innovation, offering amateur astronomers a powerful tool for exploring the universe's wonders with unparalleled clarity and precision. Renowned for its advanced coma-free (ACF) optical system, sturdy mount, and user-friendly design, the LX90 ACF is a favorite among seasoned observers and astrophotographers seeking to push the boundaries of exploration and discovery.

At the heart of the LX90 ACF telescope is its advanced optical system, which features Meade's proprietary Advanced Coma-Free (ACF) design. This revolutionary optical configuration combines a precisely figured primary mirror with a unique corrector lens to eliminate coma and produce pinpoint star images across the entire field. The result is exceptional image quality and contrast, allowing observers to discern intricate details on the surfaces of planets, galaxies, and nebulae with breathtaking clarity.

The LX90 ACF is available in a range of apertures, from 8 to 12 inches, providing ample light-gathering power to reveal faint, deep-sky objects and subtle features in the night sky. Whether observing distant galaxies millions of light-years away or exploring the intricate cloud bands of Jupiter, the LX90 ACF delivers stunning views that inspire awe and wonder in observers of all skill levels.

One of the standout features of the LX90 ACF telescope is its sturdy and reliable mount, which provides stable support and precise tracking for long-duration observations and astrophotography sessions. The telescope is equipped with Meade's AutoStar computerized GoTo system, which allows users to locate and track thousands of celestial objects automatically with the push of a button. With its built-in GPS receiver and extensive onboard database, the AutoStar system makes it easy to easily navigate the night sky and explore a wealth of celestial treasures.

In addition to its advanced optics and precise tracking capabilities, the LX90 ACF telescope features a range of user-friendly features and amenities designed to enhance the observing experience. These include a sturdy tripod with adjustable height and leveling, a built-in flip mirror for easy switching between visual and astrophotographic observing modes, and a robust dual-fork mount with integrated motors for smooth and precise movement.

The LX90 ACF telescope is also compatible with a wide range of accessories and add-ons, allowing users to customize their observing setup to suit their specific needs and preferences. Whether adding a focal reducer for wide-field astrophotography, a dew shield for dew prevention, or a solar filter for safe solar viewing, the LX90 ACF offers versatility and flexibility to accommodate various observing techniques and applications.

Observing Techniques

Observing the night sky with a telescope is a rewarding and awe-inspiring experience, but it requires patience, practice, and a solid understanding of observing techniques. Whether you're a novice astronomer or an experienced stargazer, mastering these techniques will enhance your enjoyment and appreciation of the cosmos.

Naked eye observation is the foundation of astronomy, allowing observers to identify constellations, stars, and planets visible to the unaided eye. Before turning your telescope to the sky, familiarize yourself with the celestial landscape, noting prominent landmarks and patterns. A comfortable reclining chair and a pair of binoculars can enhance the naked-eye viewing experience, revealing fainter stars and celestial objects beyond the reach of the unaided eye.

Finder scopes and star charts are indispensable tools for locating and identifying celestial objects with your telescope. Finder scopes provide a wide-field view of the sky, allowing you to pinpoint specific targets before centering them in the telescope's field of view. Star charts, whether printed or digital, serve as roadmaps of the night sky, helping you navigate the heavens and discover new wonders with each observing session. Familiarize yourself with the major constellations, asterisms, and star patterns visible from your location, using them as guideposts to navigate the celestial landscape.

Explain It To Me: Telescope

Tracking celestial objects requires precision and patience, especially when observing planets, stars, and deep-sky objects. Practice star hopping, a technique that involves moving from known stars or landmarks to your target object, using the telescope's finder scope or low-power eyepiece to navigate the sky. Once you've centered your target in the telescope's field of view, use slow and deliberate movements to track its motion across the sky, adjusting its position to keep it in view.

Consider using a motorized equatorial mount or digital tracking system to ensure steady and accurate tracking throughout your observation for objects with rapid or erratic motion, such as comets and asteroids.

By mastering these observing techniques, you'll unlock the full potential of your telescope and embark on a journey of exploration and discovery that will inspire wonder and awe for years to come.

Setting Up the Telescope

Setting up a telescope for observing can initially seem daunting, but with practice and patience, it becomes second nature. Whether you're assembling a new telescope or preparing for a night of stargazing, following these steps will ensure a smooth and successful observing session.

Assembly and alignment are the first steps in setting up your telescope. Start by carefully unpacking the telescope and its accessories, avoiding damage to delicate optical components. Follow the manufacturer's instructions to assemble the telescope, attach the optical tube to the mount, and secure any additional accessories, such as finder scopes and eyepieces. Take your time to ensure each component is aligned correctly and securely fastened, minimizing the risk of accidents or damage during use.

Balancing the mount is crucial for smooth and stable tracking of celestial objects. Adjust the position of the telescope's optical tube and any counterweights to achieve perfect balance, ensuring the mount moves smoothly and without vibration when adjusting its position. A properly balanced mount reduces strain on the telescope's motors and gears, prolonging their lifespan and improving tracking accuracy during observation.

Polar alignment is essential for observers using equatorial mounts, as it aligns the telescope's axis of rotation with the Earth's axis, allowing for precise tracking of celestial objects as they appear to move across the sky. Begin by leveling the mount using a bubble or spirit level, ensuring it is stable and securely anchored to the ground. Use a polar alignment scope or software tool to align the mount's polar axis with the celestial pole, adjusting its position until the alignment is accurate within a few arcminutes.

Once the telescope is assembled, balanced, and polar aligned, familiarize yourself with its controls and features. Practice slewing the telescope to different targets using the hand controller or keypad and experimenting with various magnifications and eyepieces to optimize your viewing experience. Use a star test or collimation tool to ensure the telescope's optics are properly aligned, fine-tuning the alignment for optimal performance.

By following these steps, you'll be ready to embark on a journey of exploration and discovery, unlocking the wonders of the cosmos with your telescope as your guide.

Common Star and Constellation Patterns

The night sky is a tapestry of stars, constellations, and celestial objects, each with unique beauty and significance. Whether you're a seasoned astronomer or a novice stargazer, understanding the common star and constellation patterns visible from your location is essential for navigating the heavens and discovering new wonders with each observing session.

Major constellations (*i.e., big dipper - Ursa Major*) serve as guideposts to the night sky, helping observers navigate the celestial landscape and locate specific stars and objects of interest. These constellations are typically divided into 88 official constellations recognized by the International Astronomical Union (IAU), each representing a specific mythological figure, animal, or object. Familiarize yourself with the major constellations visible from your location, using star charts and planetarium software to identify their distinctive shapes and patterns.

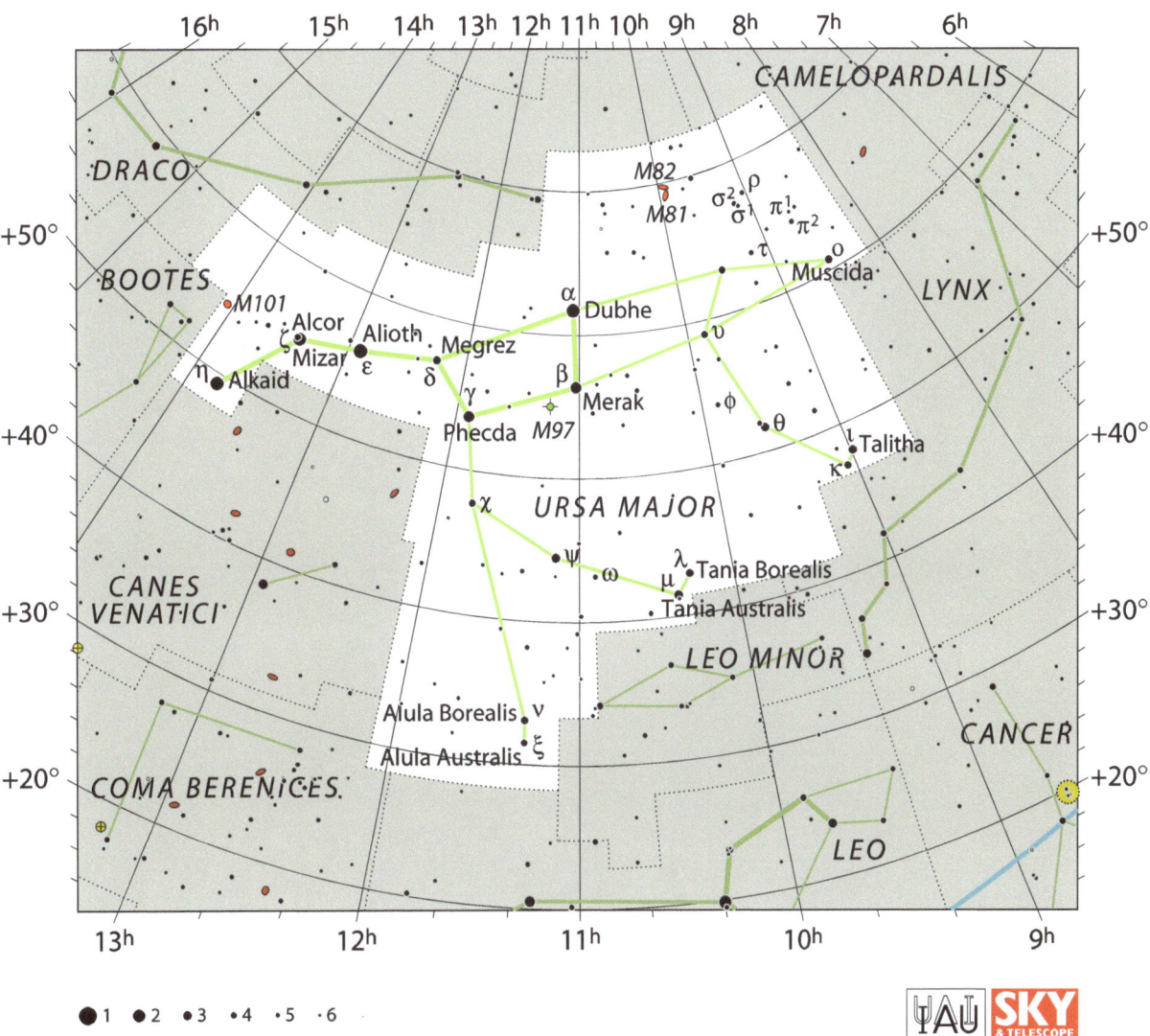

Explain It To Me: Telescope

Notable stars are scattered throughout the night sky, each with unique characteristics and properties. The brightest stars, known as first-magnitude stars, form the basis of the ancient Greek magnitude scale, which ranks stars based on their apparent brightness as seen from Earth. Among the most famous first-magnitude stars are Sirius, the Dog Star, and Alpha Centauri, the closest star system to our own. Lesser-known stars populate the sky, offering a wealth of targets for observation and exploration with telescopes and binoculars.

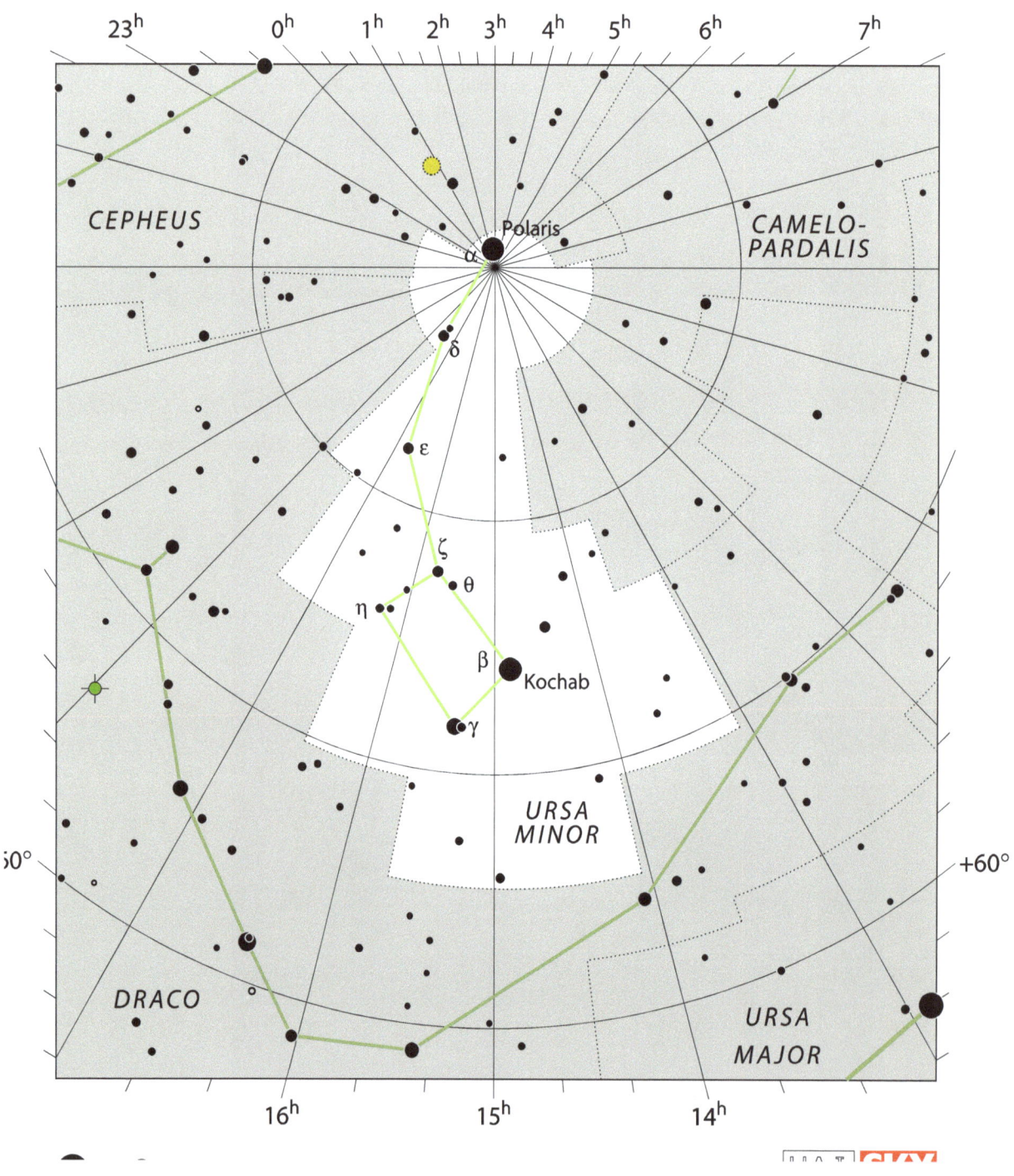

Explain It To Me: Telescope

Deep sky objects are celestial wonders beyond our solar system, including galaxies, nebulae, and star clusters. These faint and distant objects reveal the true extent of the universe's beauty and complexity, offering glimpses into distant galaxies and the birthplaces of new stars. Explore famous deep sky objects like the Orion Nebula, the Andromeda Galaxy, and the Pleiades Cluster, using telescopes and astrophotography techniques to capture their beauty and detail.

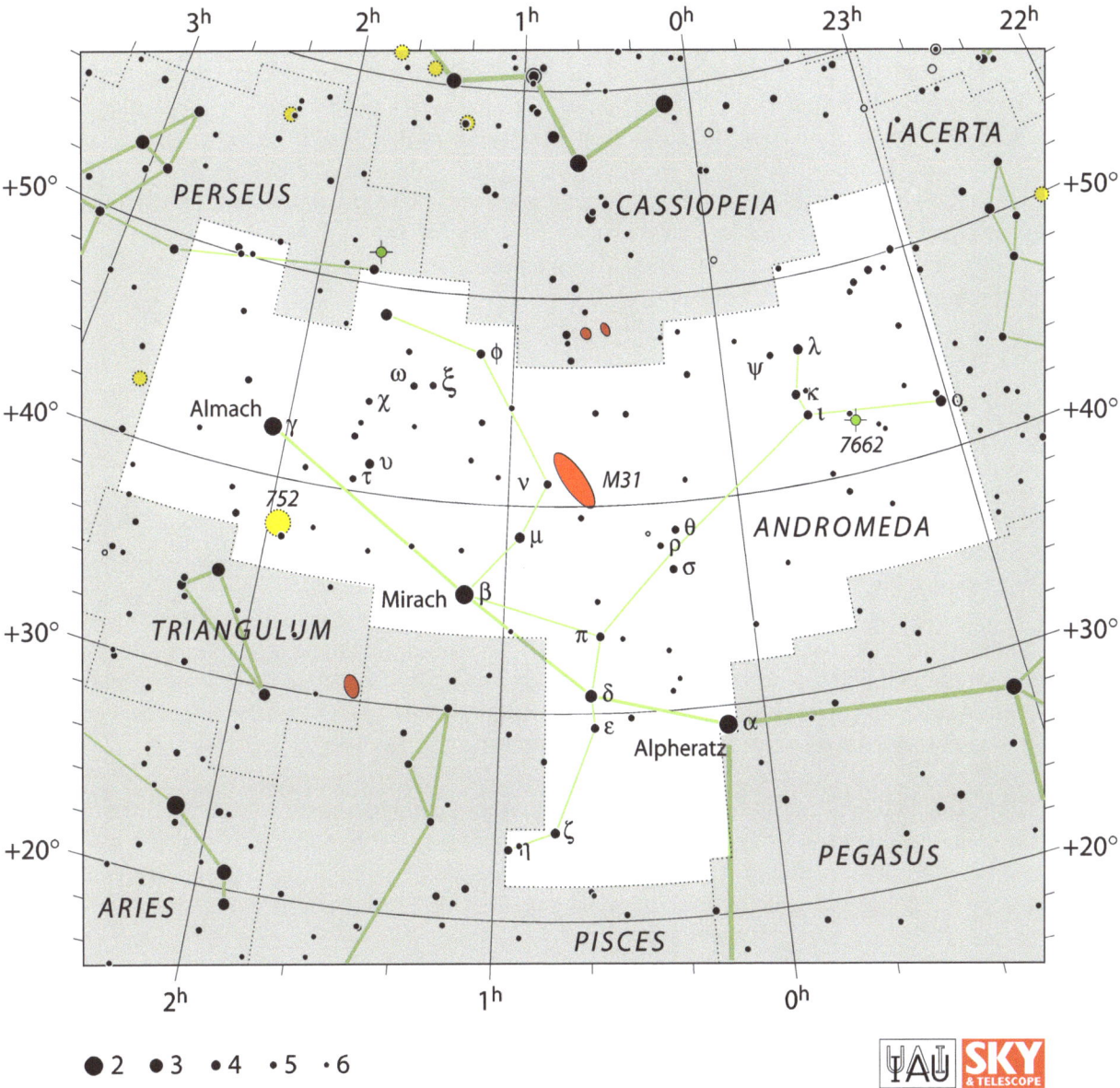

By familiarizing yourself with common star and constellation patterns, you'll gain confidence and expertise in navigating the night sky, unlocking a universe of wonder and discovery that will inspire awe and fascination for years.

Common Star and Constellation Coordinate System

Understanding the coordinate system used in astronomy is essential for locating star patterns and celestial objects in the night sky.

The primary coordinates used are Right Ascension (RA) and Declination (Dec), which provide a precise way to pinpoint the position of objects in the celestial sphere.

Right Ascension (RA) is often likened to longitude on Earth's surface, measuring eastward along the celestial equator from the vernal equinox point. The vernal equinox, where the Sun crosses the celestial equator from south to north, marks the zero point of RA. RA is typically measured in hours, minutes, and seconds, with a full circle of 360 degrees equivalent to 24 hours of Right Ascension.

Declination (Dec) is akin to latitude on Earth's surface, measuring the angular distance north or south of the celestial equator. Positive values indicate locations north of the celestial equator, while negative values indicate locations south of it. Declination is measured in degrees, with the celestial equator serving as the reference point at 0 degrees.

To use these coordinates effectively for locating star patterns, one must understand how they correspond to the celestial sphere and how they relate to the observer's position on Earth's surface. Here's a step-by-step guide to understanding and using the coordinate system:

1. **Familiarize Yourself with the Celestial Sphere**: Visualize the celestial sphere as an imaginary sphere surrounding Earth, with stars and celestial objects fixed on its surface. The celestial equator divides the celestial sphere into northern and southern hemispheres, much like Earth's equator divides the planet into northern and southern hemispheres.

2. **Understand Right Ascension**: Right Ascension is measured along the celestial equator from the vernal equinox point. Since Earth rotates once on its axis in approximately 24 hours, one hour of Right Ascension corresponds to 15 degrees of rotation. Thus, RA values increase eastward from the vernal equinox point, with each hour corresponding to 15 degrees.

3. **Learn Declination**: Declination measures the angular distance north or south of the celestial equator. Positive values indicate locations north of the celestial equator, while negative values indicate locations south of it. The maximum value for Declination is +90 degrees (the North Celestial Pole) and the minimum value is -90 degrees (the South Celestial Pole).

4. **Use Coordinates to Locate Objects**: To locate a star pattern or celestial object, use its Right Ascension and Declination coordinates to pinpoint its position on the celestial sphere. Start by identifying the general area of the sky where the object is located based on its Declination, then use its Right Ascension to pinpoint its exact position within that region.

5. **Use Reference Points and Guides**: Familiarize yourself with prominent constellations, asterisms, and bright stars that serve as reference points for navigating the night sky. By using these reference points in conjunction with RA and Dec coordinates, you can easily locate star patterns and celestial objects.

6. **Practice and Patience**: Like any skill, mastering celestial coordinates takes practice and patience. Spend time observing the night sky, familiarizing yourself with the coordinates of different objects, and gradually building your confidence in navigating the celestial sphere.

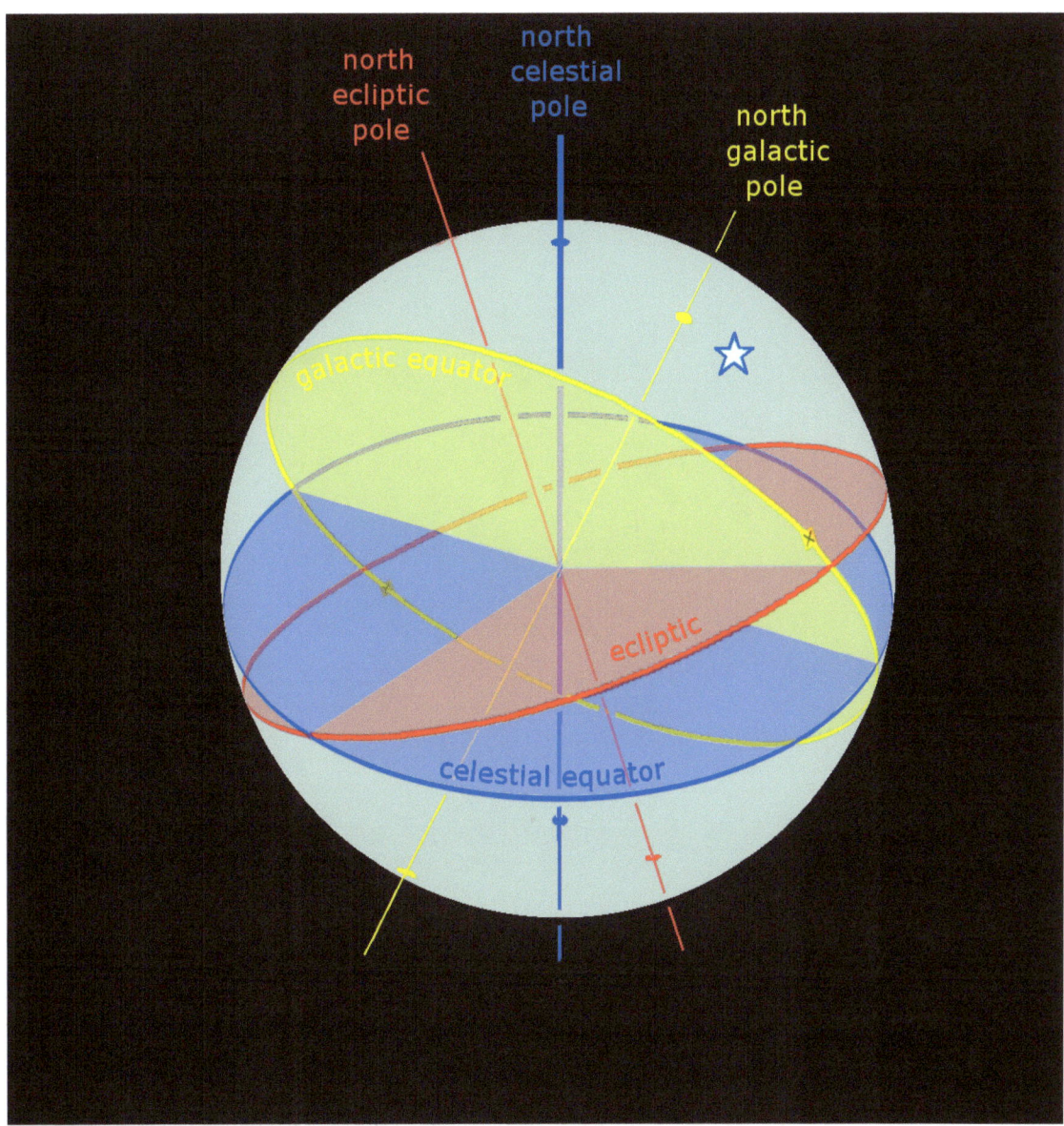

By understanding and utilizing the coordinate system effectively, amateur astronomers can unlock the wonders of the night sky and embark on a journey of discovery and exploration that transcends the boundaries of space and time. Whether observing familiar constellations or exploring distant galaxies, the celestial coordinates are invaluable tools for navigating the vast expanse of the cosmos with precision and accuracy.

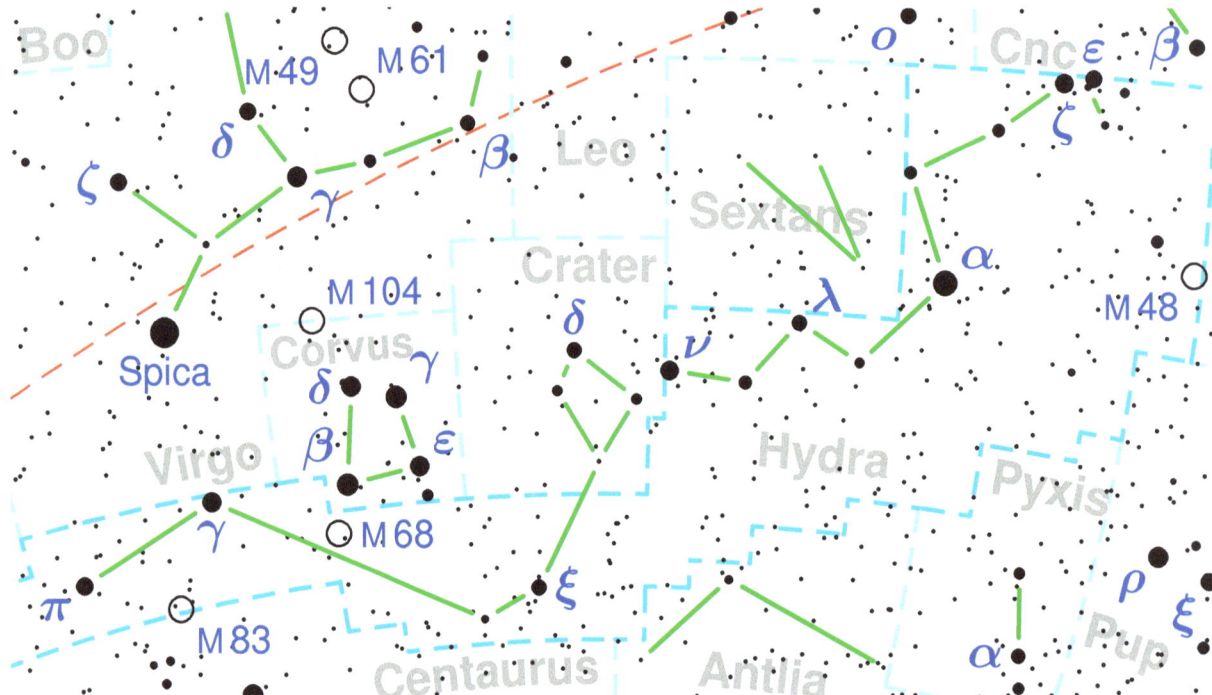

Northern Hemisphere Constellation Coordinates

Here are 20 common Northern Hemisphere star patterns, along with their approximate coordinates (i.e., right accession, declination), that can serve as guides for stargazers seeking to explore the celestial tapestry above:

1. **Ursa Major (The Great Bear):** This prominent constellation features the iconic asterism known as the Big Dipper, consisting of seven bright stars arranged in a distinctive ladle shape. Look for Ursa Major in the northern sky, with coordinates between 10h and 14h. Right Ascension (RA) and +50° to +90° Declination (Dec).

2. **Ursa Minor (The Little Bear):** Also known as the Little Dipper, this constellation contains the North Star, Polaris, which lies at the end of the handle. Ursa Minor can be found near Ursa Major, with coordinates approximately 14h and 24h RA and +60° to +90° Dec.

3. **Cassiopeia (The Queen):** Recognized for its distinctive W or M shape, Cassiopeia is a prominent constellation in the northern sky. Look for Cassiopeia between 00h and 04h RA and +60° to +90° Dec.

4. **Orion (The Hunter):** One of the most recognizable constellations, Orion features the distinctive pattern of three bright stars forming the hunter's belt, surrounded by the brilliant stars Betelgeuse and Rigel. Find Orion between 04h and 06h RA and -10° to +25° Dec.

5. **Taurus (The Bull):** Located near Orion, Taurus boasts the bright red star Aldebaran, which marks the eye of the celestial bull. The V-shaped Hyades star cluster is also a notable feature. Look for Taurus between 03h and 06h RA and +10° to +30° Dec.

6. **Leo (The Lion):** This constellation features the distinctive pattern of stars known as the Sickle, which outlines the lion's head, along with the bright star Regulus marking its heart. Find Leo between 09h and 11h RA and +10° to +40° Dec.

7. **Cygnus (The Swan):** Also known as the Northern Cross, Cygnus is a striking constellation that spans the width of the Milky Way. Look for the cross-shaped pattern of stars between 19h and 22h RA and +30° to +60° Dec.

8. **Lyra (The Lyre):** Recognized for its bright star Vega, Lyra is a small but distinctive constellation featuring a parallelogram of stars that represents the body of the lyre. Find Lyra between 18h and 20h RA and +30° to +50° Dec.

9. **Aquila (The Eagle):** Located near Cygnus, Aquila features the bright star Altair, which marks the eagle's eye. Look for Aquila between 18h and 20h RA and +00° to +30° Dec.

10. **Draco (The Dragon):** This sprawling constellation winds around the North Star, forming a serpentine shape in the northern sky. Find Draco between 15h and 22h RA and +60° to +90° Dec.

11. **Bootes (The Herdsman):** Recognized for its kite-shaped pattern of stars, Bootes is a prominent constellation in the northern sky. Look for Bootes between 13h and 15h RA and +15° to +50° Dec.

12. **Cepheus (The King)**: This constellation features a distinctive pentagonal shape of stars that represents the king's head. Find Cepheus between 20h and 02h RA and +50° to +80° Dec.

13. **Perseus (The Hero)**: Named after the legendary hero Perseus, this constellation features the bright star Algol, also known as the Demon Star. Look for Perseus between 02h and 04h RA and +25° to +60° Dec.

14. **Andromeda (The Princess)**: Home to the Andromeda Galaxy, this constellation features a distinctive "V" shape of stars that represents the princess's body. Find Andromeda between 00h and 03h RA and +20° to +50° Dec.

15. **Pegasus (The Winged Horse)**: Recognized for its large square of stars, known as the Great Square of Pegasus, this constellation is located near Andromeda. Look for Pegasus between 21h and 01h RA and +00° to +35° Dec.

16. **Gemini (The Twins)**: Featuring the bright stars Castor and Pollux, Gemini is a prominent constellation in the northern sky. Find Gemini between 06h and 08h RA and +20° to +40° Dec.

17. **Auriga (The Charioteer)**: This constellation features the bright star Capella, which marks the charioteer's head. Look for Auriga between 05h and 07h RA and +30° to +50° Dec.

18. **Pisces (The Fish)**: Recognized for its distinctive V-shaped pattern of stars, Pisces is located in the western part of the northern sky. Find Pisces between 00h and 03h RA and +00° to +30° Dec.

19. **Cancer (The Crab)**: This constellation features the open star cluster known as the Beehive Cluster and the bright star cluster Praesepe. Look for Cancer between 07h and 10h RA and +10° to +30° Dec.

20. **Aries (The Ram)**: Recognized for its distinctive triangle of stars, Aries is located in the eastern part of the northern sky. Find Aries between 01h and 03h RA and +15° to +30° Dec.

Southern Hemisphere Constellation Coordinates

Here are 20 common star patterns, along with their approximate coordinates, that can serve as guides for stargazers seeking to explore the celestial tapestry in the Southern Hemisphere:

1. **Crux (The Southern Cross)**: One of the most iconic constellations, Crux features four bright stars arranged in the shape of a cross. Look for Crux between 11h and 13h Right Ascension (RA) and -50° to -90° Declination (Dec).

2. **Centaurus**: This prominent constellation boasts the bright stars Alpha Centauri and Beta Centauri, which form the famous Southern Pointers. Centaurus can be found between 11h and 15h RA and -30° to -60° Dec.

3. **Orion (The Hunter)**: Although primarily visible from the Northern Hemisphere, Orion can also be seen from the Southern Hemisphere during certain times of the year. Look for Orion between 04h and 06h RA and -10° to +25° Dec.

4. **Carina (The Keel)**: Recognized for its distinctive shape resembling the keel of a ship, Carina is a prominent constellation in the southern sky. Find Carina between 06h and 11h RA and -30° to -60° Dec.

5. **Puppis (The Stern)**: This constellation represents the stern or poop deck of the mythical ship Argo Navis. Puppis can be found between 06h and 09h RA and -20° to -40° Dec.

6. **Vela (The Sail)**: Located near Carina, Vela represents the sail of Argo Navis. Look for Vela between 08h and 11h RA and -30° to -50° Dec.

7. **Canis Major (The Great Dog)**: Featuring the bright star Sirius, the Dog Star, Canis Major is a prominent constellation in the southern sky. Find Canis Major between 06h and 09h RA and -20° to -40° Dec.

8. **Hydra (The Water Snake)**: This sprawling constellation winds its way across the southern sky, representing the mythical serpent slain by Hercules. Hydra can be found between 08h and 15h RA and -10° to -30° Dec.

9. **Crater (The Cup)**: Located near Hydra, Crater represents the cup of the mythical god Apollo. Look for Crater between 10h and 12h RA and -10° to -20° Dec.

10. **Corvus (The Crow)**: Recognized for its distinctive quadrilateral shape, Corvus is a small but prominent constellation in the southern sky. Find Corvus between 11h and 13h RA and -10° to -25° Dec.

11. **Scorpius (The Scorpion)**: This striking constellation features the distinctive shape of a scorpion, with the bright red star Antares marking its heart. Look for Scorpius between 15h and 17h RA and -25° to -40° Dec.

12. **Sagittarius (The Archer)**: Recognized for its teapot shape, Sagittarius is a prominent constellation in the southern sky. Find Sagittarius between 17h and 20h RA and -15° to -30° Dec.

13. **Capricornus (The Sea Goat)**: This constellation features a distinctive V-shaped pattern of stars representing the head and horns of the mythical sea goat. Look for Capricornus between 20h and 22h RA and -15° to -30° Dec.

14. **Aquarius (The Water Bearer)**: Representing the water bearer of ancient mythology, Aquarius is a sprawling constellation in the southern sky. Find Aquarius between 20h and 23h RA and -10° to -30° Dec.

15. **Pisces Australis (The Southern Fish)**: Located near Aquarius, Pisces Australis represents the southern fish. Look for Pisces Australis between 22h and 01h RA and -20° to -35° Dec.

16. **Tucana (The Toucan)**: This small but distinctive constellation represents the toucan bird. Tucana can be found between 22h and 01h RA and -60° to -90° Dec.

17. **Phoenix**: Named after the mythical bird, Phoenix is a faint but recognizable constellation in the southern sky. Find Phoenix between 23h and 03h RA and -45° to -60° Dec.

18. **Eridanus (The River)**: This sprawling constellation winds its way through the southern sky, representing the mythical river Eridanus. Look for Eridanus between 02h and 06h RA and -05° to -40° Dec.

19. **Octans (The Octant)**: This small but distinctive constellation represents the octant, a navigational instrument sailors use. Octans can be found between 20h and 24h RA and -75° to -90° Dec.

20. **Musca (The Fly)**: Located near Crux, Musca is a small but recognizable constellation in the southern sky. Find Musca between 10h and 13h RA and -60° to -75° Dec.

These common star patterns guide stargazers seeking to explore the wonders of the night sky, offering a wealth of celestial treasures waiting to be discovered and admired. Whether you're a seasoned observer or a beginner taking your first steps into astronomy, these constellations provide a captivating glimpse into the beauty and mystery of the cosmos.

Observing Planets and Other Celestial Bodies

Observing planets and other celestial bodies through a telescope is a thrilling and rewarding experience, offering glimpses into distant worlds and the mysteries of the cosmos. Whether you're a beginner or an experienced astronomer, mastering the techniques and strategies for observing planets, stars, and deep-sky objects will enhance your enjoyment and appreciation of the night sky.

Observing planets requires careful planning and timing, as their positions and appearances change yearly. Begin by consulting a planetarium app or online almanac to determine which planets are visible from your location and their approximate positions in the sky. Planetary observing sessions are best conducted during evenings when the planets are high in the sky and free from atmospheric distortion caused by turbulence and light pollution.

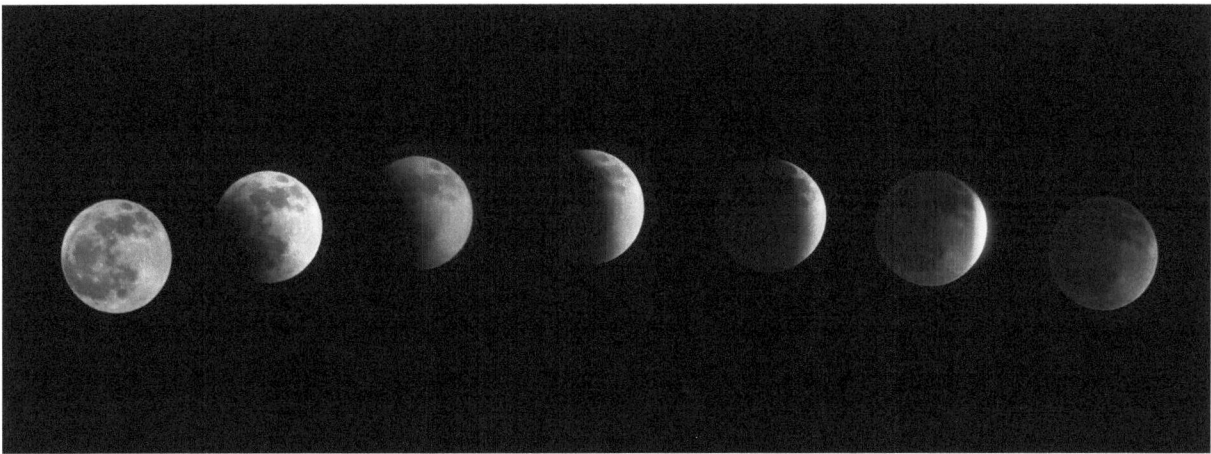

When observing planets through a telescope, use moderate to high magnifications to reveal finer details in their surface features and cloud patterns. Start with low-power eyepieces to locate the earth and center it in the telescope's field of view, then gradually increase magnification using higher-power eyepieces to enhance contrast and resolution. Experiment with different filters, such as color filters and polarizing filters, to enhance surface features and improve the visibility of atmospheric details like dust storms and cloud bands.

In addition to planets, the night sky is teeming with other celestial objects, including stars, nebulae, and galaxies. Deep sky objects, such as the Orion Nebula, the Andromeda Galaxy, and the Pleiades Cluster, offer stunning views through telescopes and binoculars, revealing intricate structures and delicate filaments invisible to the naked eye. Plan observing sessions around the moon's phases and the positions of significant constellations using star charts and astronomy apps to guide your night sky exploration.

By mastering the techniques and strategies for observing planets and other celestial bodies, you'll unlock the full potential of your telescope and embark on a journey of discovery and exploration that will inspire awe and wonder for years to come.

Observation Tools & Accessories

Observing the night sky with a telescope is an exhilarating experience that can be enhanced using various tools and accessories to aid navigation, observation planning, and celestial object identification. From smartphone apps to practical accessories, these tools can significantly improve the enjoyment and effectiveness of telescope observation for amateur astronomers of all skill levels.

Let's explore some of the most commonly used tools and accessories:

1. **Stellarium and SkySafari**: Stellarium and SkySafari are popular planetarium apps for iOS and Android devices. These apps provide detailed sky maps and celestial databases, allowing users to quickly identify stars, planets, constellations, and deep-sky objects. They also feature advanced features such as telescope control, night mode, and time-lapse animations, making them invaluable tools for planning, observing sessions, and exploring the night sky.

2. **Star Charts and Atlases**: Physical star charts and atlases, such as "Sky & Telescope's Pocket Sky Atlas" or the "Cambridge Star Atlas," are essential for navigating the night sky and locating celestial objects. These pocket-sized reference books contain detailed maps of the stars, constellations, and deep-sky objects visible from different locations and times of the year, making them ideal companions for telescope observation sessions in the field.

3. **Red Flashlight**: A red flashlight is essential for preserving night vision during telescope observation sessions. Unlike white light, which can disrupt the eye's ability to adapt to low-light conditions, red light has minimal impact on night vision. It allows observers to read star charts, adjust telescope settings, and navigate dark observing sites without affecting their ability to see faint celestial objects.

4. **Telescope Filters**: Telescope filters, such as moon filters, light pollution filters, and narrowband filters, can enhance the contrast and visibility of celestial objects by reducing glare, enhancing contrast, and blocking unwanted light pollution or atmospheric distortion. These filters can significantly improve the quality of observations, especially when observing bright objects like the moon or planets from urban or light-polluted areas.

5. **Eyepiece and Filter Cases**: Carrying cases or organizers for eyepieces and filters are essential for keeping telescope accessories organized and protected during transportation and storage. These cases typically feature padded compartments or foam inserts to securely hold eyepieces, filters, and other accessories, preventing them from getting damaged or misplaced.

6. **Telescope Control Software**: Telescope control software, such as StellariumScope or EQMOD for equatorial mounts, allows users to control their telescope remotely from a computer or mobile device. This software provides advanced features for precise tracking, automatic slewing, and image capture, making locating and tracking celestial objects easier, especially when using complex or computerized telescopes.

7. **Collimation Tools**: Collimation tools, such as collimation caps, lasers, or Cheshire eyepieces, are used to align and adjust the optical components of a telescope, ensuring that the mirrors or lenses are correctly aligned for optimal performance. Proper collimation is essential for achieving sharp, high-quality images and accurate telescope observations.

8. **Observing Chair**: A chair or stool with adjustable height is essential for comfortable telescope observation sessions, especially during long-duration observations or astrophotography sessions. A comfortable observing position can reduce fatigue and strain on the neck and back, allowing observers to enjoy the night sky for extended periods without discomfort.

9. **Power Supply**: A reliable power supply, such as a portable power pack or rechargeable battery, is essential for powering electronic accessories like telescopes, cameras, and laptop computers during observing sessions in the field. It's critical to ensure that the power supply is fully charged and capable of providing sufficient power for the duration of the observation session.

10. **Notebook and Pen**: Keeping a notebook or observation log is a great way to record observations, sketch celestial objects, and jot down notes about observation conditions, equipment settings, and memorable observations. A waterproof notebook and pen are recommended for observing sessions in the field, where weather conditions may be unpredictable.

By utilizing these tools and accessories, amateur astronomers can enhance their telescope observation sessions, improve their observing skills, and unlock the full potential of their telescopes for exploring the wonders of the night sky. Whether observing from the backyard, a dark sky site, or a remote observing location, these practical tools can significantly enhance the enjoyment and effectiveness of telescope observation for stargazers of all ages and experience levels.

Light Pollution and Dark Sky Sites

Light pollution poses a significant challenge to astronomers and stargazers, obscuring the natural beauty of the night sky and diminishing the visibility of stars, planets, and other celestial objects. Understanding the causes and effects of light pollution and strategies for mitigating its impact is essential for preserving and enjoying the wonders of the cosmos.

Light pollution is caused by the excessive and inefficient use of artificial light, including streetlights, outdoor lighting, and industrial sources. The glare and skyglow produced by these sources scatter light into the atmosphere, creating a luminous haze obscuring faint stars and celestial objects. Light pollution not only diminishes the aesthetic beauty of the night sky but also disrupts ecosystems, affects human health, and wastes energy and resources.

However, the recent proliferation of low Earth orbit (LEO) satellites has raised new concerns about exacerbating light pollution and preserving dark skies. Companies like SpaceX, Amazon, and OneWeb have launched thousands of small satellites into LEO as part of their plans to provide global broadband internet coverage. While these satellite constellations have the potential to bring high-speed internet access to underserved areas around the world, they also pose a significant threat to the visibility of the night sky.

To combat light pollution and preserve the natural beauty of the night sky, astronomers and environmentalists advocate for establishing dark sky sites—areas free from artificial light pollution where the night sky remains pristine and unspoiled. Dark sky sites are typically located in remote or rural areas with minimal light pollution, such as national parks, nature reserves, and designated dark sky preserves. These sites offer unparalleled opportunities for stargazing, astrophotography, and scientific research, providing a glimpse into the majesty and grandeur of the universe.

Finding dark sky sites requires research and planning, as not all locations offer optimal conditions for observing the night sky. Consult dark sky maps and atlases to identify nearby dark sky sites and their accessibility, considering weather conditions, local regulations, and seasonal variations in light pollution levels. When visiting dark sky sites, respect the natural environment and observe any rules or guidelines established to protect the area from light pollution and other environmental degradation.

By supporting efforts to combat light pollution and preserve dark sky sites, astronomers and stargazers can ensure future generations can experience the beauty and wonder of the night sky in all its glory. Whether observing from your backyard or venturing to a remote dark sky site, take time to appreciate the natural splendor of the cosmos and the profound sense of awe and wonder it inspires.

Space Telescopes

Space-based telescopes, such as the iconic Hubble Space Telescope, represent the pinnacle of astronomical observation, pushing the boundaries of our understanding of the universe and delivering breathtaking images of celestial objects with unprecedented clarity and detail. These instruments, situated above Earth's atmosphere, offer unique advantages over their ground-based counterparts. They enable scientists to study the cosmos across various wavelengths and unlock the universe's secrets.

The Hubble Space Telescope, launched in 1990, is one of the most iconic and revolutionary space telescopes ever built. Named after the pioneering astronomer Edwin Hubble, the telescope has transformed our understanding of the cosmos and provided invaluable insights into fundamental astrophysical processes, from the formation of galaxies and stars to the nature of dark matter and dark energy.

Explain It To Me: Telescope

The primary mission of space telescopes like Hubble is to observe the universe across a wide range of wavelengths, from ultraviolet and visible light to infrared and beyond. By operating above Earth's atmosphere, which can distort and absorb incoming light, space telescopes can capture images with unparalleled clarity and precision, free from the effects of atmospheric turbulence and light pollution.

One of the critical advantages of space telescopes is their ability to observe celestial objects continuously without interruption from Earth's day-night cycle or weather conditions. This allows astronomers to conduct long-duration observations of distant galaxies, nebulae, and other astronomical phenomena, providing a wealth of data that would be difficult or impossible to obtain with ground-based telescopes. In addition to their observational capabilities, space telescopes like Hubble are equipped with advanced instrumentation and detectors that enable scientists to study the universe in unprecedented detail. These instruments can analyze celestial objects' composition, temperature, and motion, providing crucial insights into their physical properties and histories.

Furthermore, space telescopes are not limited by the constraints of Earth's surface, allowing them to observe the universe across a broad range of wavelengths inaccessible to ground-based telescopes. For example, Hubble's ultraviolet and infrared detectors have revealed hidden features of the cosmos, such as the presence of young stars and the formation of planetary systems that would be invisible to the human eye or traditional optical telescopes.

Despite their many advantages, space telescopes also face unique challenges and limitations. Operating in a harsh space environment, these instruments are subject to cosmic radiation, micrometeoroid impacts, and degradation of sensitive components over time. Maintenance and repairs, such as the famous Hubble Servicing Missions, are often necessary to ensure space telescopes' continued operation and longevity.

Another challenge is the finite lifespan of space telescopes, which are typically limited by the availability of onboard fuel, the degradation of critical components, or the depletion of scientific instruments. However, advancements in space technology and engineering have enabled the development of next-generation space telescopes, such as the James Webb Space Telescope (JWST), which promises to build upon the legacy of Hubble and expand our understanding of the universe even further.

In contrast to space telescopes, amateur telescopes are typically ground-based instruments that rely on optical lenses or mirrors to collect and focus light from celestial objects. While amateur telescopes may lack the sensitivity and precision of space-based instruments, they offer enthusiasts the opportunity to explore the night sky from their backyard and observe various celestial phenomena, from the moon and planets to distant galaxies and nebulae. They also come in various designs and configurations, ranging from simple refracting or reflecting telescopes to more complex compound or computerized models. Despite their limitations, amateur telescopes play a crucial role in engaging the public in the wonders of astronomy and inspiring the next generation of scientists and explorers.

Space telescopes like the Hubble Space Telescope represent marvels of modern engineering and scientific ingenuity, enabling astronomers to peer deeper into the cosmos than ever before and unravel the mysteries of the universe. While they differ from amateur telescopes in their complexity, capabilities, and operating environment, both instruments contribute valuable insights to our understanding of the cosmos and inspire wonder and curiosity in observers of all ages and backgrounds.

Maintenance and Care

Proper maintenance and care are essential for ensuring the longevity and performance of your telescope, allowing you to enjoy years of rewarding observation and exploration. Whether you're a beginner or an experienced astronomer, following these tips and guidelines will help keep your telescope in optimal condition and ready for action whenever inspiration strikes.

Cleaning optics is a crucial aspect of telescope maintenance, as dust, dirt, and debris can accumulate on the telescope's lenses and mirrors, degrading image quality and clarity. Use a soft, lint-free cloth or brush to gently remove any surface dust or particles from the optics, avoiding scratching or damaging delicate optical coatings. For stubborn stains or residues, use a mild optical cleaning solution or isopropyl alcohol diluted with distilled water, apply it sparingly, and avoid contact with sensitive components such as electronic controls and focus mechanisms.

Collimation is another essential maintenance task that ensures the telescope's optics are properly aligned, allowing for crisp, high-resolution images of celestial objects. Use a collimation tool or star test to check the alignment of the telescope's mirrors and lenses, making adjustments to correct any misalignment or distortion. Regular collimation improves image quality and sharpness, especially when observing planets and other high-resolution targets.

Storage tips are essential for protecting your telescope from damage and deterioration when not in use. Store the telescope in a dry, climate-controlled environment away from direct sunlight, extreme temperatures, and humidity, which can cause warping, corrosion, and mold growth. Use protective covers or cases to shield the telescope from dust, moisture, and physical damage during storage and transportation, securing any loose components and accessories to prevent loss or damage.

Following these maintenance and care tips, ensure your telescope remains in peak condition and ready for action whenever the mood strikes. Whether observing from your backyard or venturing to a remote dark sky site, a well-maintained telescope will provide years of rewarding observation and exploration, unlocking the wonders of the cosmos and inspiring awe and wonder in all who gaze upon it.

Practical Uses for the Telescope

Telescopes are versatile tools with many practical applications beyond recreational stargazing and amateur astronomy. Whether you're a student, educator, scientist, or enthusiast, the telescope offers opportunities for exploration, discovery, and learning that transcend the boundaries of space and time.

Amateur astrophotography is one of the most popular and rewarding telescope uses, allowing enthusiasts to capture stunning night sky images and share them with the world. Whether photographing distant galaxies, colorful nebulae, or the planets of our solar system, astrophotography offers a creative outlet for expressing your passion for astronomy and sharing the beauty of the cosmos with others.

Educational outreach is another valuable application of telescopes, providing students and the public with hands-on opportunities to explore the universe's wonders and learn about the principles of astronomy and astrophysics. Whether hosting a star party, leading a classroom demonstration, or participating in a public observing event, telescopes can inspire curiosity, wonder, and a lifelong love of learning in people of all ages and backgrounds.

Scientific research contributions are also possible with telescopes, as amateur astronomers and citizen scientists contribute valuable data and observations to professional research projects and scientific studies. Whether you're monitoring variable stars, tracking asteroids, or searching for exoplanets, telescopes offer opportunities for amateur astronomers to make meaningful contributions to our understanding of the cosmos and the processes that shape it.

Whether exploring the night sky from your backyard or venturing to a remote dark sky site, the telescope offers endless opportunities for discovery, exploration, and learning. From amateur astrophotography to educational outreach and scientific research, telescopes empower enthusiasts of all ages and backgrounds to unlock the mysteries of the universe and share their passion for astronomy with the world.

Key Terms

Acknowledging the importance of key terms and concepts in astronomy and telescope usage is essential for fostering understanding and communication among enthusiasts and educators. Whether you're a beginner exploring the night sky for the first time or an experienced astronomer seeking to deepen your knowledge, familiarizing yourself with these key terms will enhance your enjoyment and appreciation of the cosmos.

1. **Accessories:** Additional components and attachments used to enhance the telescope's performance and versatility, including Barlow lenses, filters, focal reducers, and finderscopes.

2. **Amateur astrophotography**: The practice of capturing images of celestial objects using telescopes, specialized cameras, software, and techniques. Amateur astrophotographers contribute valuable data and observations to scientific research projects and public outreach efforts.

3. **Aperture**: The diameter of the primary optical component of a telescope, whether it's a lens or a mirror. Aperture determines the telescope's light-gathering ability and resolution, with larger apertures collecting more light and revealing finer details in celestial objects.

4. **Arcminute**: This, denoted by the symbol ('), is a unit of angular measurement commonly used in astronomy, cartography, and navigation. It is defined as one-sixtieth (1/60) of a degree of arc. Just as there are 60 minutes in an hour in timekeeping, there are 60 arcminutes in a degree of arc. Arcminutes are used to specify small angles, particularly in contexts where precision is essential, such as determining the apparent size of celestial objects in the sky.

5. **Astrophotography:** The practice of capturing and recording images of celestial objects and phenomena using specialized cameras and equipment. It combines elements of astronomy and photography to produce stunning images of stars, planets, galaxies, nebulae, and other astronomical subjects. To capture pictures of celestial objects, astrophotographers use a variety of specialized cameras, telescopes, and accessories, including CCD (charge-coupled device) or CMOS (complementary metal-oxide-semiconductor) cameras, telescopes with tracking mounts, autoguiders, and filters.

6. **Citizen science**: The participation of amateur astronomers and enthusiasts in scientific research projects and studies, contributing valuable data and observations to professional research efforts.

7. **Collimation**: The process of aligning the telescope's optical components, such as mirrors and lenses, to ensure optimal image quality and sharpness. Collimation is essential for maintaining the telescope's performance and accuracy during observation.

8. **Dark sky sites**: Areas free from artificial light pollution where the night sky remains pristine and unspoiled, offering unparalleled opportunities for stargazing, astrophotography, and scientific research.

9. **Eyepieces**: The optical components used to magnify and view celestial objects through the telescope. Eyepieces come in various focal lengths and designs, each providing a different magnification and field of view.

10. **Focal length**: The distance from the primary optical component to the focal point, where light rays converge to form an image. The focal length determines the telescope's magnification and field of view, with longer focal lengths providing higher magnifications and narrower fields.

11. **Mount type**: The mount used to support and control the telescope, whether an alt-azimuth or equatorial mount. Mount type affects the telescope's stability, tracking accuracy, and suitability for different observing techniques and applications.

12. **Star hopping**: A technique used to navigate the night sky by moving from known stars or landmarks to your target object, using the telescope's finder scope or low-power eyepiece to guide your exploration.

13. **Space telescope**: An astronomical instrument designed to observe celestial objects beyond Earth's atmosphere. Unlike ground-based telescopes, which are limited by atmospheric distortions and light pollution, space telescopes operate in the vacuum of space, allowing them to capture images with unparalleled clarity and precision across a wide range of wavelengths.

By familiarizing yourself with these key terms and concepts, you'll enhance your understanding and appreciation of astronomy and telescope usage, empowering you to explore the wonders of the cosmos with confidence and enthusiasm.

Acknowledgments

The publication of this comprehensive guide on the use of telescopes would not have been possible without the contributions and support of numerous individuals and organizations. We extend our heartfelt thanks to the astronomers, educators, and enthusiasts who generously shared their expertise and passion for astronomy, enriching the content and scope of this publication.

Special thanks to Celestron for their dedication to advancing the field of amateur astronomy and providing enthusiasts with high-quality telescopes and accessories that inspire wonder and discovery. We are grateful for their commitment to innovation, education, and outreach, which have made the joys of stargazing and exploration accessible to people worldwide.

We also acknowledge the invaluable contributions of amateur astronomers, educators, and citizen scientists who promote the appreciation and understanding of the night sky through public outreach, educational programs, and scientific research initiatives. Their enthusiasm, dedication, and tireless efforts inspire curiosity, wonder, and a lifelong love of learning in people of all ages and backgrounds.

Finally, we express our gratitude to the readers and enthusiasts who have embraced the wonders of the cosmos and embarked on a journey of discovery and exploration with telescopes as their guide. May this publication inspire you to look up, explore, and discover the beauty and majesty of the universe in all its glory.

Conclusion

In conclusion, telescopes are potent tools that open windows to the universe, revealing the beauty and majesty of the cosmos in all its glory. From the earliest observations of Galileo to the modern wonders of the Hubble Space Telescope, telescopes have expanded our understanding of the cosmos and inspired generations of astronomers, scientists, and enthusiasts to explore the wonders of the night sky.

By understanding the history, types, and practical uses of telescopes, as well as mastering observing techniques and strategies, enthusiasts can embark on a journey of discovery and exploration that will inspire awe and wonder for years to come. Whether you're a beginner exploring the night sky for the first time or an experienced astronomer seeking to push the boundaries of discovery, the telescope offers endless opportunities for exploration, learning, and inspiration.

May this publication serve as a valuable resource and companion on your journey through the cosmos, empowering you to unlock the mysteries of the universe and share your passion for astronomy with the world. Remember to look up, explore, and discover the wonders of the night sky with telescopes as your guide, and may the beauty and majesty of the cosmos inspire awe and wonder in all who gaze upon it.

Credits

	Publication References
a.	Resources and information were obtained from organizations such as the International Astronomical Union (IAU), the American Astronomical Society (AAS), and amateur astronomy clubs provide valuable information on observing techniques, equipment recommendations, and practical tips for stargazing.
b.	Aperture diagram by Cmglee; data on holes in mirrors provided by an anonymous user from IP 71.41.210.146 - Own work, CC BY-SA 3.0, https://commons.wikimedia.org/w/index.php?curid=33613161
c.	Focal length diagram by Henrik - Own work, GFDL, https://commons.wikimedia.org/w/index.php?curid=3698171
d.	Celestial diagram by Tfr000 (talk) 16:32, 25 June 2012 (UTC) - Own work, CC BY-SA 3.0, https://commons.wikimedia.org/w/index.php?curid=20028939
e.	Diagram on Andrometer constellation: https://commons.wikimedia.org/wiki/File:Andromeda_IAU.svg#/media/File:Andromeda_IAU.svg
f.	Diagram on Ursa Minor constellation: https://commons.wikimedia.org/wiki/File:Andromeda_IAU.svg#/media/File:Andromeda_IAU.svg
g.	Diagram on Ursa Major constellation: https://commons.wikimedia.org/wiki/File:Ursa_Major_IAU.svg#/media/File:Ursa_Major_IAU.svg
h.	Diagram on Hydra: https://commons.wikimedia.org/wiki/File:Ursa_Major_IAU.svg#/media/File:Ursa_Major_IAU.svg
i.	Image of the Hubble Telescope - by Ruffnax (Crew of STS-125) - http://catalog.archives.gov/OpaAPI/media/23486741/content/stillpix/255-sts/STS125/STS125_ESC_JPG/255-STS-s125e011848.jpg, Public Domain, https://commons.wikimedia.org/w/index.php?curid=6826183
j.	The Magic of Stargazing - Leisure Travel Vans. https://leisurevans.com/blog/the-magic-of-stargazing/
k.	How To Buy A Telescope - Telescope Nerd. https://www.telescopenerd.com/telescope-astronomy-articles/how-to-buy-a-telescope.htm
l.	STARGAZER S-109XY Telescope Astronomical Refractor – Clarity Scopes. https://clarity-scopes.com/products/stargazer-s-109xy-telescope-astronomical-refractor Meade Polaris 127mm German Equatorial Reflector - 216005 I – Khan Scope Centre. https://khanscope.com/products/meade-polaris-127mm-german-equatorial-reflector-216005 Choose a telescope - Deep Sky. https://deepsky2000.net/choose-a-telescope-3/ Nediyedath, Avinash S., et al. "24 New Citizen Science Light Curves of WASP-12b and Updated Ephemeris by Combining with ETD and ExoClock Datasets." 2023, http://arxiv.org/abs/2306.17473.
m.	The Impact of Light Pollution on Astronomy and The Study of the Universe. https://aniprocs.com/index.php/blog/29-the-dark-side-of-lighting/85-the-impact-of-light-pollution-on-astronomy-and-the-study-of-the-universe
n.	The Space Telescope That Could Find a Second Earth	Air & Space Magazine| Smithsonian Magazine. https://www.smithsonianmag.com/air-space-magazine/picture-planet-180977302/
o.	Some photos have been provided by Murray Campbell, Daniel Sinoca, Jeremy Tomas, Corina Rainer, Claude Schwarz, Gage Smith, Jared Craig, Igal Ness, Patrick Hendry and Alexander Andrews on Unsplash.com.
p.	Various other photos provided by Laney & Associates, LLC.
q.	Some photos were purchased from VectorStock.
r.	Some photos were provided from Wikipedia under the Public Domain license.